ADDITION & SUBTRACTION
COLOR BY NUMBERS

Please consider writing a review!
Just visit: wizolearning.com/review

Copyright 2020. Wizo Learning.
All Rights Reserved.

No part of this book may be reproduced or transmitted in any form or by any means, electronic or mechanical, including photocopying, recording or by any other form without written permission from the publisher.

Have questions? We want to hear from you!
Email us at: support@wizolearning.com

ISBN: 978-1-951806-35-4

FREE BONUS

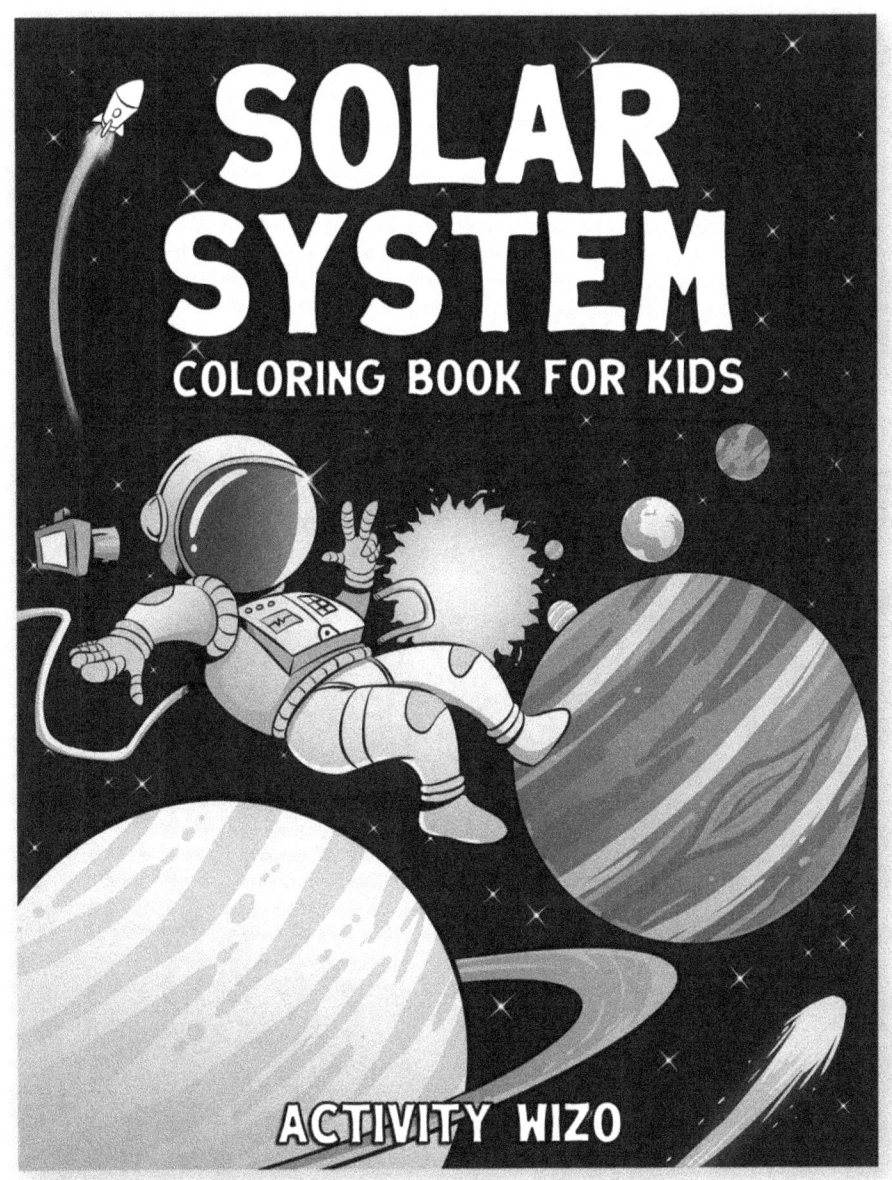

Just flip to the end of the book
to get the link!

② BLUE   ③ YELLOW   ⑤ ORANGE   ⑦ GREEN

① YELLOW   ② BROWN   ③ RED   ④ ORANGE

(2) ORANGE   (10) YELLOW   (8) PINK   (7) BROWN

⑤ PINK ⑦ RED ⑨ ORANGE ⑩ BLUE

(3) ORANGE   (4) YELLOW   (7) GRAY   (9) BROWN

④ GREEN ⑤ BROWN ⑦ BLUE ⑨ PURPLE

① PURPLE   ② ORANGE   ③ YELLOW   ④ BLUE

(6) PURPLE   (2) ORANGE   (4) BROWN

(7) RED   (8) BLUE

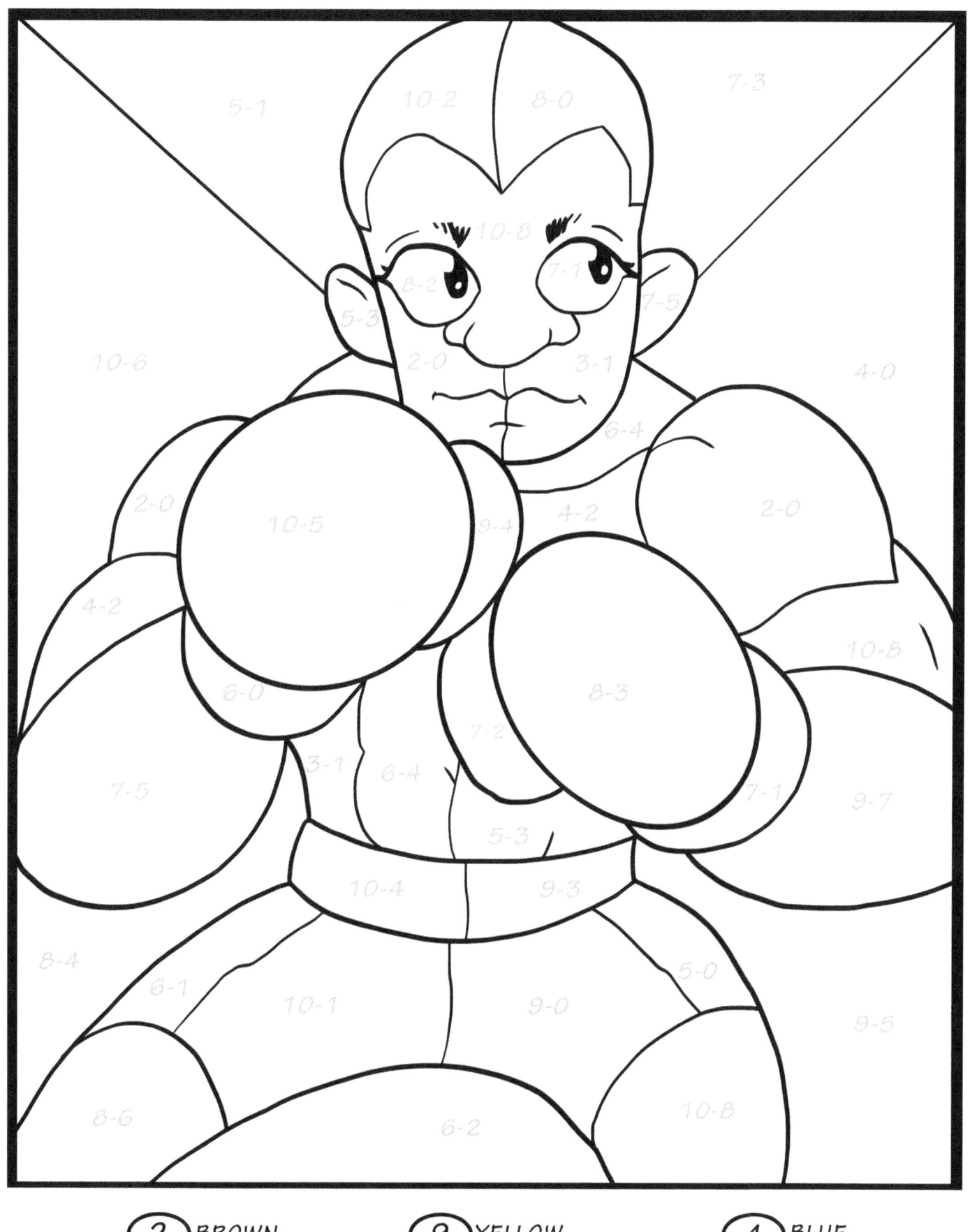

FREE BONUS

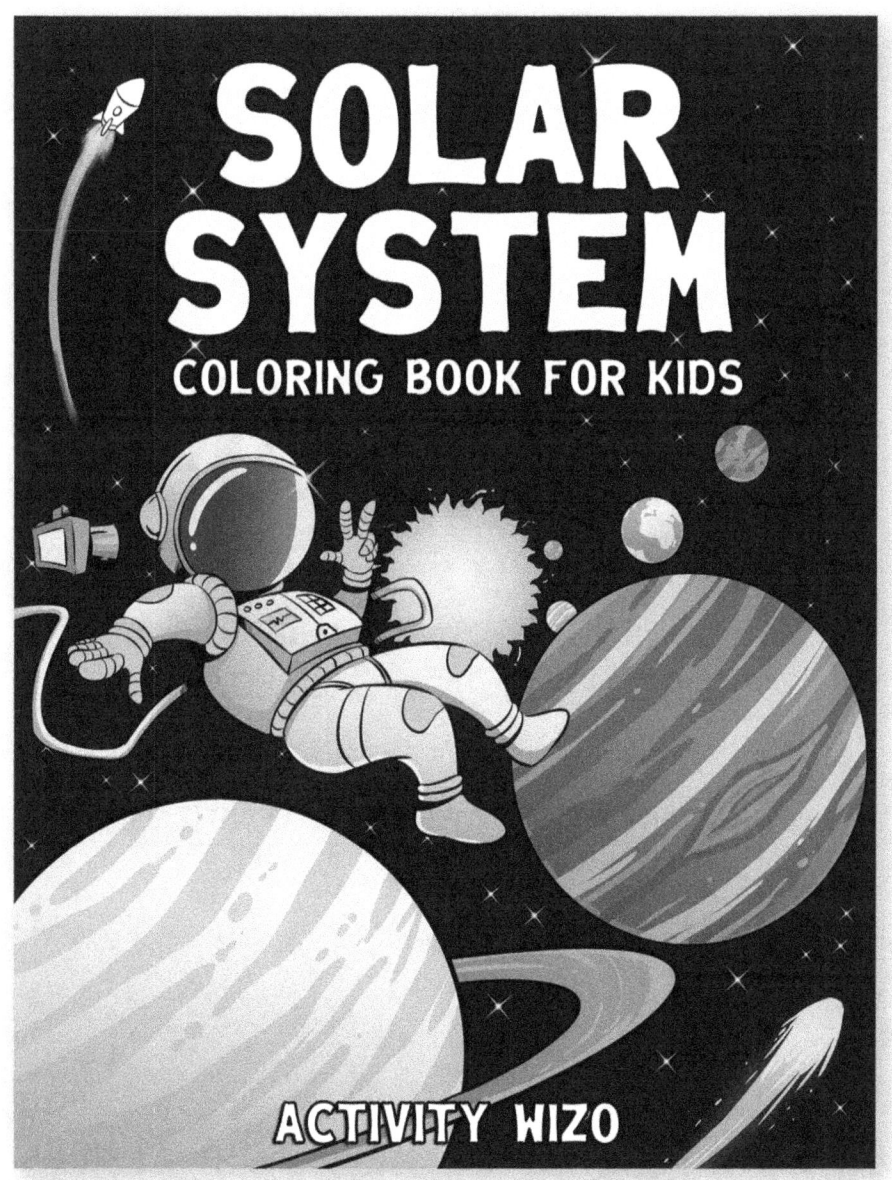

Get This FREE Bonus Now!

Just go to: activitywizo.com/free

THANK YOU!

Have questions? We want to hear from you!
Email us at: support@activitywizo.com

Please consider writing a review!
Just visit: activitywizo.com/review

www.ingramcontent.com/pod-product-compliance
Lightning Source LLC
Chambersburg PA
CBHW081754100526
44592CB00015B/2435